Make a Sundial

This booklet has been prepared by members of the Education Group of the British Sundial Society.

David Brown -------------- Physics teacher, Kingswood School, Bath.
Paul Higgins ------------- Mathematics teacher, France Hill School, Camberley.
Michael Maltin ---------- Airline Pilot (retd), Stroud.
Pat Insole-Moore ------ Primary School teacher (retd.)
David Nicholls ---------- Lecturer in 3D Art and Design, Colchester.
Dave Pawley ----------- Tower clock maker, Newbury.
Graham Stapleton ----- Administrator, Iveagh Bequest, Kenwood.
Jane Walker ------------- Mathematics teacher, France Hill School, Camberley.
Peter Walker ------------ Engineer (retd.), Sandhurst.

The Education Group wishes to thank France Hill School, Camberley and Kingswood School, Bath for the use of computing and other facilities.

How to use the book

The background section and the activities which go with it should be tackled first.
The construction projects may be done in any order though some are easier than others.
Teacher's notes are provided on blue sheets for easy reference. Although many of the dials described can be made in a few minutes, it is essential to read through the sheets and to try out the constructions before embarking on a class project.
A glossary has been included.
Pupils' pages (yellow and white) may be photo-copied for classroom use.

All the dials in this book are designed for use in the northern hemisphere between latitudes 49 degrees and 59 degrees.

© British Sundial Society 1991   Printed at France Hill School, Camberley.   ISBN 0 9518404 0 1

Sundials

Preface

The inauguration of the British Sundial Society in 1990 coincided with the introduction of the National Curriculum in schools which included the following requirements:-
(children) ' should investigate the use of a sundial as a means of observing / noting / charting the passage of time .'

In 1992 the Programme of Study for Science AT4 paragraph v states:-
Pupils should track the path of the sun using safe procedures such as a shadow stick or sundial.

Since the best way to understand how a sundial works is to make one, we have received a great many requests from teachers for practical instructions for constructing a sundial in the classroom. This book is our response to such requests.

Our approach has been deliberately simple as the most important need was for an easily made dial which would work. A list of appropriate texts is included for those requiring a more rigorous treatment of the subject.
In addition to the dials which can be constructed by individual pupils in a classroom we have included, in the teachers' notes, sections on finding the direction of south and on the construction of a playground, analemmatic, dial. These are more suitable for group projects.
Similarly, the section on constructing an equatorial dial from a bicycle wheel and a broomstick is intended to produce a classroom demonstration model rather than a project for individual pupils.

Sundials are works of art as well as of science, they therefore provide an ideal starting point for cross curricular studies. Science, mathematics and geography are all involved in the construction projects and there are obvious links with history. Art and poetry are also involved; a 'dial hunt' in your area may reveal both ancient and modern examples of the dialists skill and ingenuity. Some have added embellishments such as points of the compass or signs of the zodiac and appropriate 'words of wisdom' on the nature of time are often included.

We hope the booklet will prove both interesting and enjoyable and would welcome your comments on its usefulness in the classroom.

The British Sundial Society has as one of its aims the compilation of a comprehensive catalogue of the sundials of the British Isles. This is a mammoth task which may not be completed for some time but a stamped addressed envelope (plus £1) to the Secretary (see BSS addresses) will bring a list of the dials catalogued so far which can be visited in your area.

Notes   Sundials

Contents

	page
Background activities (yellow pages)	
Be back in time	4
1 - shadows	5
2 - make a shadow stick	6
3 - moving shadows	7
4 - keeping track of the sun	8
5 - a dial to colour	10
Construction projects (white pages)	
1 - A horizontal dial	14
2 - A horizontal dial from a matchbox	18
3 - An equatorial dial	20
4 - An equatorial dial from a washing-up bottle	22
5 - A polar dial from a matchbox	24
6 - A polar dial made from wood or card	26
7 - A sundial time conversion calculator	30
Further activities (yellow pages)	
1 - Sundial time and clock time	32
2 - Dial hunting	36
Group projects (white pages)	
1 - A bicycle wheel and broomstick dial	38
2 - An analemmatic dial	44
General notes for teachers (blue pages)	50
Notes on activities 1 to 5	54
Notes on projects 1 to 6	55
Finding south	61
Tables for simple dials	65
Glossary	66
Booklist	68
British Sundial Society addresses	70
Photographs	12,13,17,43

Sundials

background

Be back in time

Imagine you lived in the iron age, or that you live today in a remote part of the world. You don't go to school but your job is to look after your family's flock of goats. You have to take them out to graze every day and bring them back at night. You have no watch and neither has anyone else. There are no public clocks, no radio or television, no telephones.
How would you know when to head for home so that you'll be safely back in the village before dark ?

On a holiday you can go out to play with your friends but your mother says, "*Be back in time for dinner.*" How will you know when to come back ?.
What things would change about you as the day wears on ?
The sun would climb higher in the sky and it would get warmer so you could tell when it was midday - well, more or less!

The lengths of shadows would also change. Your mother might say "*Come back when the shadow of the pine tree reaches the river bank.*" All very well if you are playing near the pine tree !

On one day you could find the length of your own shadow at the time when the shadow of the pine tree reaches the river.

How would you measure the length of your own shadow ?

A better idea might be to use the shadow of a stick which you could carry with you. Then by sticking it in the ground and noticing the length of its shadow you would know when it was time to go home.

Sundials
background

Activity 1 - shadows

Working in pairs, make a chart to show the length of your shadow at each hour of the day.

Fill in the blanks in this chart.

When my shadow is _____ cm long I set off for school.

When my shadow is _____ cm long we start school.

When my shadow is _____ cm long we have playtime.

When my shadow is _____ cm long we have dinner.

When my shadow is _____ cm long we go home.

When my shadow is _____ cm long I go to bed.

Sundials
background

Activity 2 - make a shadow stick

You will need

> A cotton reel
> A large sheet of white paper
> A pencil
> A table close to a sunny window

Push the pencil into the cotton reel. Make sure it is upright.
Place the cotton reel on the paper on a flat surface near a sunny window.
Draw round the pencil's shadow.
Do not move the cotton reel.
Come back 5 or 6 times during the day and draw round the shadow each time.

Questions

1) What have you noticed about the shadow's position?
   Can you explain why?

2) What have you noticed about the shadow's length?
   Can you explain why?

Activity 3 - moving shadows

Do you have a netball post in your school playground?

Record the time and the length of the post's shadow at 3 or 4 different times throughout the school day.

Measure the length of the shadow again at the same times every day for one week. You could make out a table to record the shadow lengths for each time on each day.

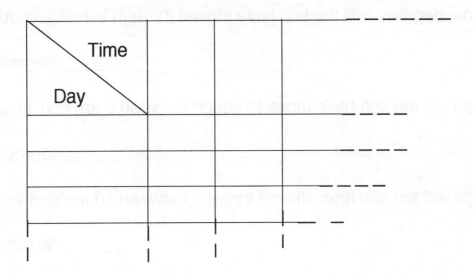

Make a chalk mark on the playground to show where the end of the shadow is when the bell goes for the end of playtime.

On the next day leave your watch at home and see if you can tell when the bell should go.

Sundials
background

Activity 4 - keeping track of the sun

The sun appears to move round the earth, making one complete circuit of 360 degrees in 24 hours. So it moves through 15 degrees in each hour (because 360 / 24 = 15).

In 2 hours the sun will have moved 30 degrees

In 3 hours the sun will have moved 45 degrees ... and so on.

How many degrees will the sun have moved through in half an hour?

_____degrees

What angle will the sun have moved through between 9 am and 11 am?

_____degrees

What angle will the sun have moved through between 10 am and 2pm?

_____degrees

What angle will the sun have moved through between 8 am and 5 pm?

_____degrees

By constructing a very simple device we can use this regular movement of the sun to show the time of day. Such a device is called a sundial.

Sundials
background

Did you notice the shadow of your post moving around ?

Draw a picture to show what you noticed.
Show the shadow at two or three different times during the day.

Write a few sentences to describe what you noticed.

Did you notice anything happening to the **length** of the shadow of your post during the day?
Write a few sentences to say what you noticed.

Try to explain what you have seen - think about what is causing the shadow to move and change in length.

Sundials
background

Activity 5 - a dial to colour

1. Colour the dial plate shown opposite and the gnomon shown at the bottom of this page.

2. Write a few words about time in the space below the sun on the dial plate.

   For example: TIME FLIES

   or: TIME AND TIDE WAIT FOR NO MAN

   Or you can make up your own.

3. Cut out the plate and gnomon and stick them on to card.
4. Cut a slot in the plate where the dotted line is drawn.
5. Glue the gnomon tabs to the *underside of* the dial plate to the left and right of the slot.
6. When the glue has dried make sure that the gnomon is exactly upright.

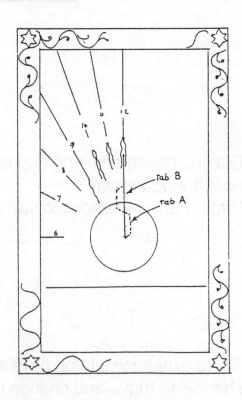

7. Put your dial on a flat surface in a sunny position with the gnomon lying exactly in the north-south direction and the upper end towards north.

> You can now read the sun time from your dial.

Sundials
background

horizontal sundial plate. lat 52°.

Sundials

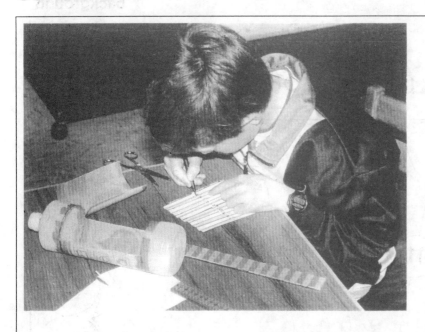

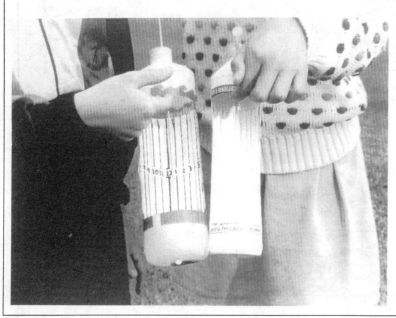

Pupils at Finmere School making sundials.

Sundials

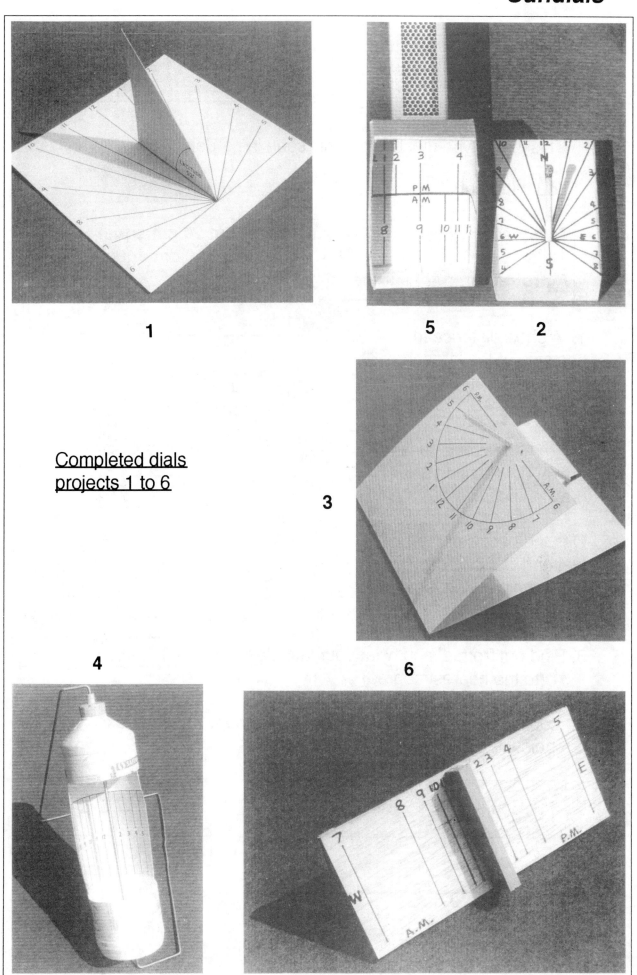

Completed dials
projects 1 to 6

Photographs on these 2 pages by Francis Scott

Sundials
project 1

A horizontal dial

This is the type most commonly seen in gardens.

You will need

**Two pieces of card or plywood, one about 20 cm square
the other about 10 cm square
A protractor, ruler and pencil**

1. A short distance from one side of the larger square, and parallel to it, draw a line AB.

2. From the centre O of AB and at right angles to it, draw a line OC.

O is the centre of the dial and OC the line along which the gnomon will lie.

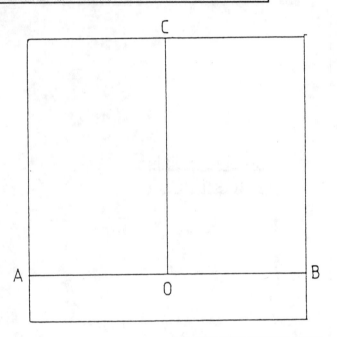

3. Find out from a map what your latitude is (to the nearest degree)

Latitude =

4. Look at the table on p. 65 and from it copy into the boxes below the angles for the hour lines

Hours	12	11 & 1	10 & 2	9 & 3	8 & 4	7 & 5	6
Angles	0						90

Sundials

project 1

5. Use your protractor to draw lines from O at the angles you have put in the table. The hours of morning are to the left of C and the afternoon hours are to the right.

6. Write the hours next to the lines.

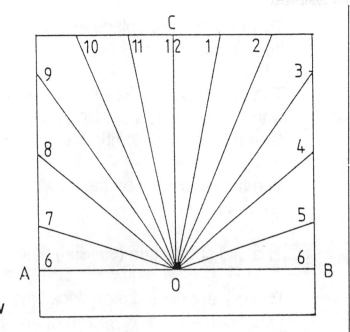

   Your drawing should now look like that opposite.

7. On the smaller piece of board draw a right-angled triangle ODE. The angle DOE must be equal to your latitude. Triangle ODE is the gnomon of your sundial.

8. Cut out the gnomon ODE together with the tab (only if card is used - a plywood gnomon will need some small pieces of quadrant beading or similar to support it.)

9. Stick the gnomon ODE on to the dial plate with O of the gnomon exactly at O of the plate and the line OE exactly along the line OC on the plate. The gnomon must be exactly at right angles to the dial plate.

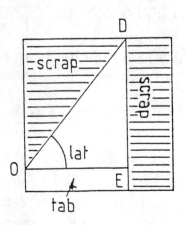

Sundials   Notes
project 1

10. **To use your dial,** place it on a horizontal surface in a sunny open position.

    The gnomon must lie exactly north-south,
    with the upper end pointing towards north.
    (see p.61 on how to find the north-south direction).

    The time is shown by the shadow of the edge OD of the gnomon.

    The dial shows sun (sundial) time.

    To work out mean time (clock time) you will need to make the usual corrections as shown on p.35 or use the sundial time conversion calculator shown on p.30.

    Example: suppose that your sundial has been correctly set up on a horizontal surface with its gnomon in the north-south direction.

Your longitude is 2 degrees west.
The date is July 1st.

Using page 35 (or pages 32 to 34 for more explanation) you can see that you need to

   (a) add 1 hour for British Summer Time .... + 1h 00m

   (b) add 8 minutes for 2 deg. of longitude west ... ... +   08m

   (c) add 3 minutes from the time conversion graph ..+   03m
                                                              ─────
                                     total correction .... +1h 11m

   So whatever the sundial time was on this day you would need to add 1 hour and 11 minutes to it to get it to agree with your clock.
   (The sundial time conversion calculator on p. 30 does the same calculation but makes it easier for you.)

Sundials

armillary sphere

vertical declining dial

horizontal memorial dial

stained glass window dial

Things to look for when dial hunting (see p 36)

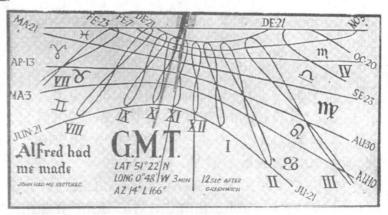

an unusual wall dial

Sundials
project 2

A horizontal dial from a matchbox

This sundial takes only a few minutes to make.

You will need

> **An empty matchbox**
> **One used matchstick**
> **A pen or pencil**
> **A ruler**

1) Take out the tray of the matchbox and turn it upside down.

For steps 2 to 6 see Fig 1.

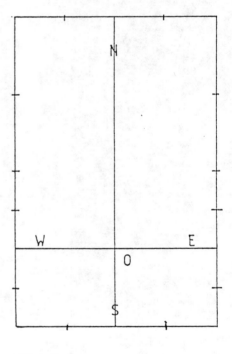

Fig 1

2) Divide each edge into 4 equal parts.
3) Join the centre points of the two short sides and mark **N** and **S**.
4) On the long sides, join the first quarter division points nearest to **S** and mark **E** and **W**.
5) Mark **O** where the two lines cross.
6) Divide in half again the two bottom quarters on each long side above and below the **E W** line.

Sundials

project 2

For steps 7 and 8 see Fig. 2.

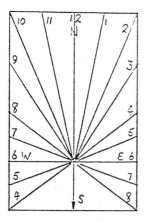

Actual size

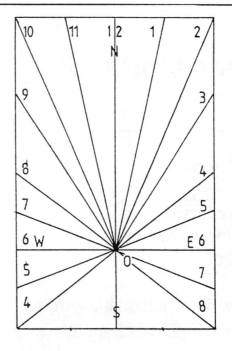

Fig 2

7) From **O** draw lines out to all of the points around the edges and to the corners of the box.
8) Number the lines as shown in Fig. 2

For steps 9 and 10 see Fig 3.

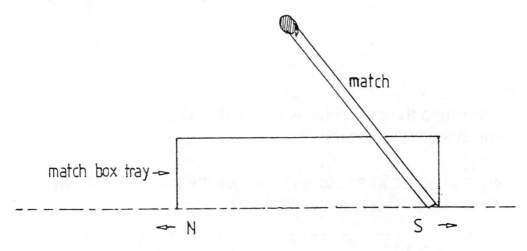

Fig 3

9) Make a hole at **O** and push the matchstick through. Slope the matchstick so that its end rests against the end of the tray as shown.
10) Place the dial on a flat surface in sunlight so that the **N S** line is in the north - south direction with the matchstick pointing to the north.

To compare the sundial with your watch see the section on **sundial time and clock time** (p. 32).

Sundials

project 3

An equatorial dial

You will need

A piece of card
A drinking straw
A protractor

1) Draw lines across the card as shown in fig. 1
   Mark the points **O** and **C**.

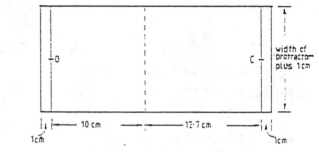

Fig 1

2) Place the centre of the protractor on **O** and draw round it. Make a pin hole through **O**, turn the card over and place the centre of the protractor on the pin hole. Draw round the protractor again so that you have two semicircles back to back.

3) Mark 15 degree intervals and number the hours, as shown in Figs. 2 and 3, on both semicircles.

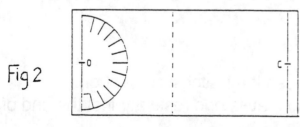

Fig 2

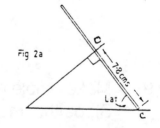

Fig 2a

4) Score and fold the card along the dotted line.

5) Make a mark on the straw 7.8 cm from one end.

6) Enlarge the pin hole at **O** and push the straw through until the mark is level with the card (Fig. 2a).

Sundials
project 3

The straw must make a right angle with the card.

7) Cut a tab at **C** slightly wider than the straw (see Fig. 3a). Push the tab into the end of the straw to hold it firmly.

8) The straw forms the **gnomon**. Its shadow will fall on the hour lines on the top of the dial in summer and on the underside in winter.

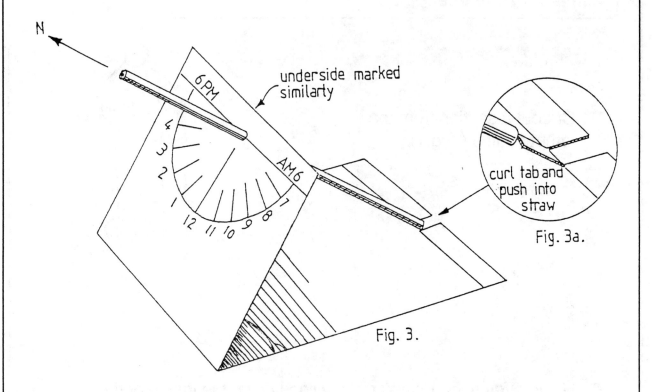

Fig. 3.

Fig. 3a.

You can colour your dial and write a motto underneath.

To compare the sundial with your watch see the section on **sundial time and clock time** (p.32).

Sundials

project 4

An equatorial dial from a washing-up liquid bottle

You will need

> An empty washing-up liquid bottle
> A knitting needle or barbecue stick or coat hanger
> Scissors
> A fine tipped marker pen

1) Cut out a section from the empty bottle (Fig.1).

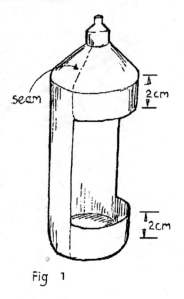

Fig 1

2) Using a felt tipped marker, draw lines down the inside of the bottle as follows:-   (Fig. 2)
   First mark the centre line, 12.
   Mark the 6 am and 6 pm lines at the edges.
   Mark 9 midway between 12 and 6 am.
   Mark 3 midway between 12 and 6 pm.
   Divide the spaces between all these marks into three and label the rest of the hour lines (all of these divisions can be made by eye).

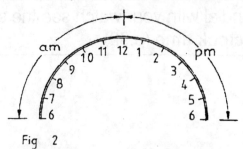

Fig 2

Sundials
project 4

3) Insert the knitting needle or barbecue stick or a straight piece of metal from the coathanger, through the top of the bottle and through the centre of the base.
   This knitting needle is now called the **gnomon**.

4) Hang the bottle from your finger as shown in Fig. 3 so that the gnomon lies in the north - south direction.

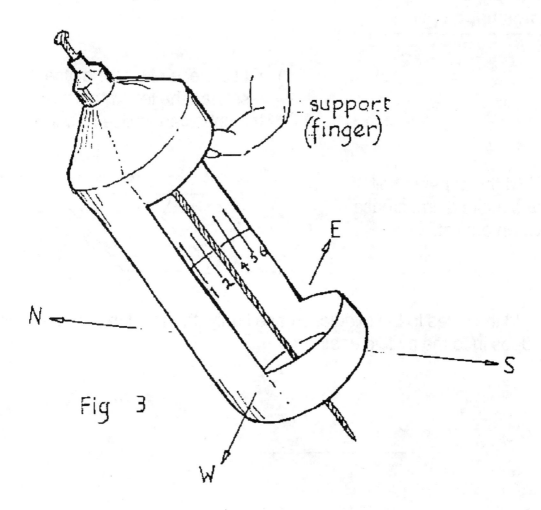

Fig 3

The shadow of the gnomon will fall on the hour lines to show the time by the sun.

To compare the sundial with your watch see the section on **sundial time and clock time** (p.32).

Sundials

project 5

A polar dial from a matchbox

You will need

> An empty matchbox
> Sticky paper or
> Plain paper and glue
> A fine tipped pen

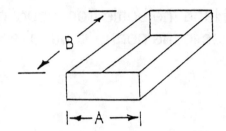

1) Cut a piece of paper a little less than the length of the matchbox and twice as wide.

2) Fold the paper in half and make a firm crease. Open out flat.

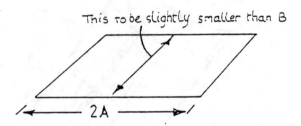

3) Fold the sides to the middle, crease firmly, then fold the creased edges in to the middle again.

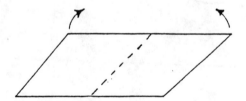

The paper will now be divided into 8 equal parts with 7 creases.

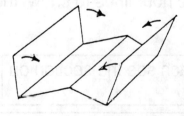 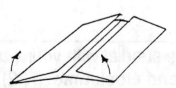 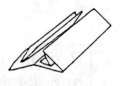

Sundials

project 5

4) Open out flat.
Fold the paper in half and make one crease across the middle in the opposite direction.
Write **pm** above and **am** below this line.

5) Use a fine pen to draw along the creases and number the lines.

		12	2	3 p|m		5	6
				a|m			
6	7			9	10	12	

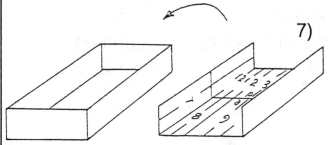

6) Draw a line midway between **12** and **2 pm** and on it mark **1pm** and **8 am**.
Draw a line midway between **10 am** and **12** and on it mark **4 pm** and **11 am**.

7) Stick the paper to the inside of the matchbox tray.

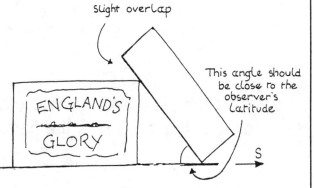

8) Set up the dial by placing the outer part of the matchbox in a north - south direction and rest the tray against it at about 52 degrees.

Slight overlap

This angle should be close to the observer's latitude.

The shadow of the side of the box will fall along the hour lines.

In the morning read the lower **am** scale.

In the afternoon read the upper **pm** scale.

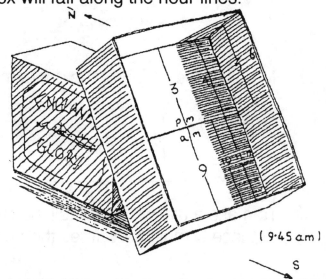

(9·45 am)

To compare the sundial time with your watch see the section on **sundial time and clock time** (p.32).

25

Sundials

project 6

A polar dial made from wood

This sort of dial was used in Ancient Egypt to check their water clocks. There is a drawing in the Science Museum, London, showing this being done.

You will need

> **2 pieces of wood** (see step 3)
> **1 piece of dowel rod**
> **A drill (size to suit dowel)**
> **Wood glue , or hammer and nails**
> **A calculator with a tan key**

1) Mark the centre line across the **dial plate.**
   Mark the centre of the bottom edge at the ends of the **gnomon**.

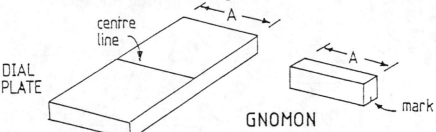

2) Keeping the marks in line, fix the gnomon across the centre of the dial plate using the glue or the hammer and nails.

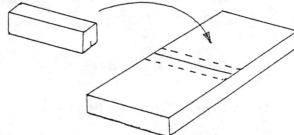

3) Measure the height of the gnomon **H**.

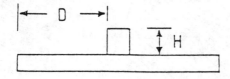

To make a dial which will tell the time from 7 am to 5 pm the distance **D** must be 4 times the height **H**.

Sundials
project 6

4) Fill in the height of the gnomon in all the empty boxes to the left of the = sign.

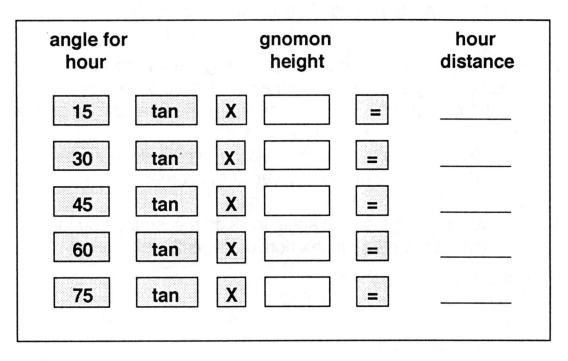

Press the calculator keys in the order shown for each row and write the result in the space on the right.

Example

If the height of your gnomon is 45 mm the first row would be :-

| 15 | tan | X | 45 | = | 12.1 mm |

The answers give you the distances of the **hour lines** which are to be drawn on the dial plate.

5) Draw the hour lines on the dial plate measuring outwards from each side of the gnomon.

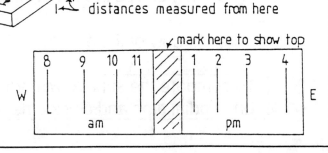

Sundials

project 6

6) Mark **E, W, am** and **pm**.
   Make a mark on the edge to show which is the top.

7) **The dial plate must be tilted at an angle equal to your latitude.**
   Find your angle of latitude from a map (about 51 degrees in southern England). This angle can be measured using a protractor or it can be drawn on a piece of card:-
   Measure 100 mm along one edge and 'h' along the other edge. Join the marks and cut off the corner.
   (See the table below for values of 'h'.)
   When using the card angle the side measuring 100 mm should always be in the horizontal position.

8) Drill a hole for the dowel.
   Glue in place sticking straight out.

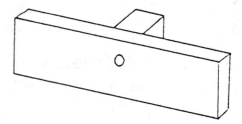

latitude (degrees)	h (mm)
49	115
50	119
51	123
52	128
53	133
54	138
55	143
56	148
57	154
58	160
59	166

9) Cut the end of the dowel until the card will just fit as shown.

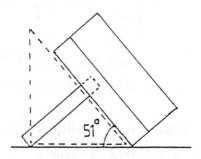

10) Place the dial in a sunny position with its ends pointing east - west.
    The gnomon lies in the north - south direction.

    The shadow of the gnomon will fall on the hour lines to tell the time.

    To compare the sundial with your watch see the section on sundial time and clock time (p.32).

Sundials
project 6

Alternative method for polar dial using card instead of dowel

Follow steps 1 to 7 on pages 26 to 28, then:-

9) Make up two more pieces of card with an angle equal to your angle of latitude.
   You can do this by drawing round the first piece of card

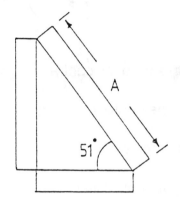

*Make **A** equal to the **width** of the dial plate*

Add small tabs as shown before cutting out.

10) Bend the tabs up on one piece and down on the other piece.
    Stick together with small pieces of sticky paper to make two supports.

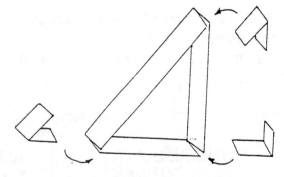

12) Stick the cards to the back of the dial plate at each end.

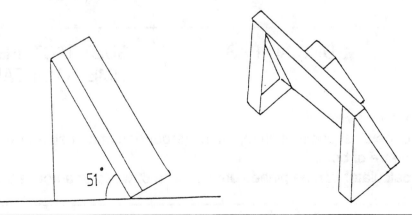

Sundials
project 7

Sundial time conversion calculator

You will need
- A sheet of A4 size card
- A sharp knife or razor blade
- A ruler and a soft pencil

1) Copy the two diagrams on the facing page on to the card or use the back cover of this book.

2) Cut out the slit below the centre of the smaller circle.

3) Draw a line between the two arrowed dots to cross this slit at its centre.

4) Find your longitude from an ordnance survey map to the nearest quarter of a degree.
   Note whether you are east or west of the Greenwich meridian.

5) With a ruler parallel to the cut out slit, measure from your pencil line a distance in millimetres equal to 4 X your longitude.
   Measure downwards if you are west of Greenwich,
   upwards, towards the centre, if you are east of Greenwich.
   From this point draw a dotted line parallel to the other dotted lines.
   This is the new zero or **datum line.**
   Rub out the pencil line you drew first.

6) From the datum line, mark off a scale alongside the slit, both upwards and downwards every 10 mm.
   Mark the datum line **0** and the other scale marks **10, 20,**------------ in both directions.

7) Write on the smaller circle as shown below.

   SUNDIAL SLOW   ADD TO SUNDIAL TIME
                        TO OBTAIN CLOCK TIME

   datum line --

   SUNDIAL FAST   SUBTRACT FROM SUNDIAL
                        TIME TO OBTAIN CLOCK TIME

8) Cut out the two circles.
   Place the smaller circle on top of the larger one and push a drawing pin through the centre of both.
   The calculator can be pinned onto a piece of wood or a notice board.

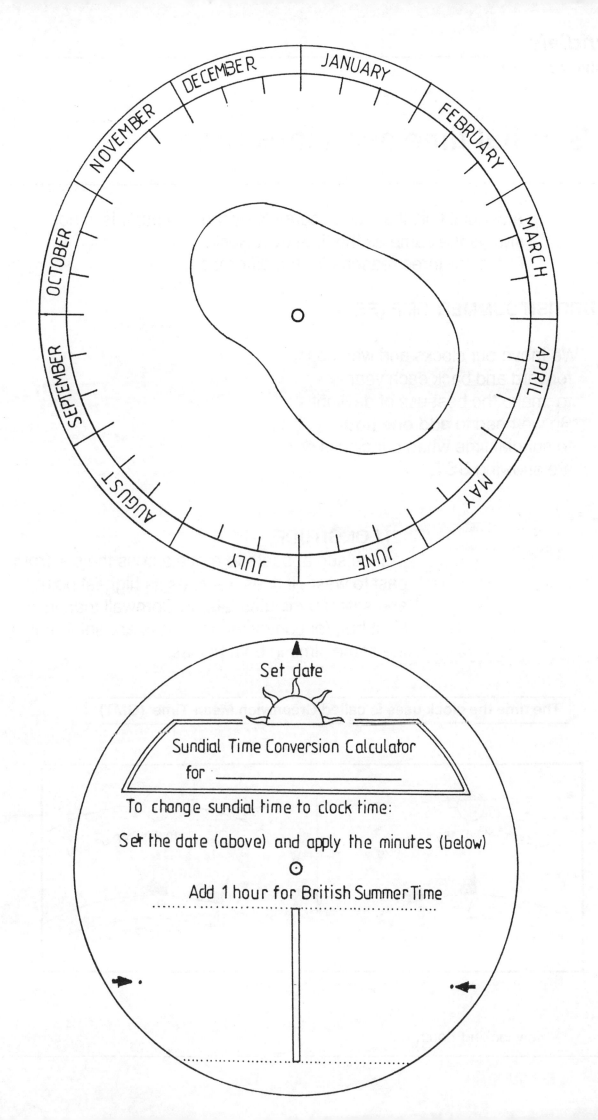

Sundials
further activity 1

Sundial time and clock time

A sundial tells the time accurately by the sun but this is not always the same as the time by a clock.
There are three reasons for the difference:-

1) BRITISH SUMMER TIME (BST)

We move our clocks and watches forward and back each year to make the best use of daylight. So we need to **add one hour** to sundial time when our clocks are showing **BST.**

2) LONGITUDE

As the sun appears to move across the sky from east to west, it rises, reaches its highest point, and sets, 20 minutes later in Cornwall than in Kent but, for convenience, clocks are set to tell the same time in both places.

The time the clock uses is called Greenwich Mean Time (GMT)

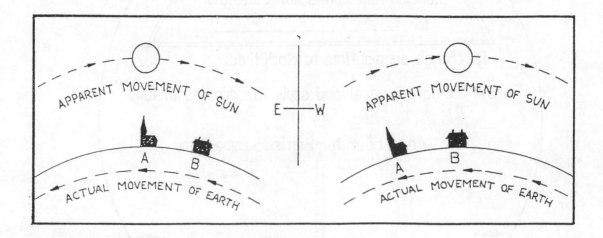

View looking South.

Sundials
further activity 1

To correct for longitude :-

longitude	correction to sundial time to obtain clock time .
every degree west of Greenwich	**add 4 minutes**
every degree east of Greenwich	**subtract 4 minutes**

3) TIME CONVERSION GRAPH

Clock time is based on the idea that the earth travels round the sun at a steady speed but in fact its speed varies slightly at different times of the year because its orbit is not a true circle but is slightly elliptical.
Also, the earth is tilted on its axis causing the apparent speed of the sun across the earth to vary with the seasons. So some 'sun' (solar) days are longer than others but 'clock' days are all the same length.
To correct for this use the time conversion graph below.

(Or use the sundial conversion calculator on page 31)

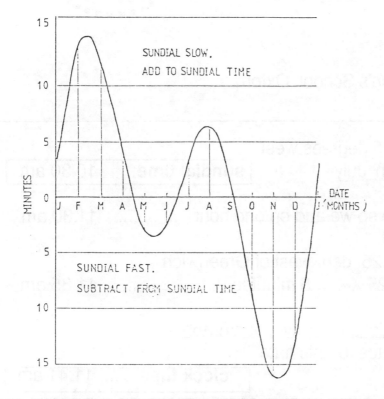

Time conversion graph for sundials
Based on the equation of time

Sundials
further activity 1

Activity - finding clock time from sundial time

An example is given first so that you can check to see whether you have taken the right steps.
You may find this flow-chart helpful.

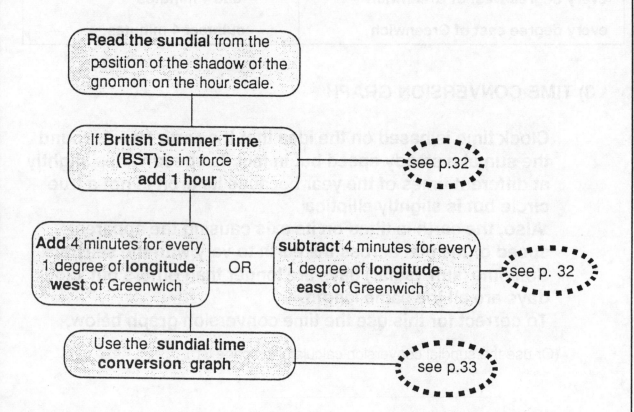

Example for St. John's School, Oxford

> **Longitude**: 1.25 degrees west
> **Date**: 17th July   **sundial time**:   10.30 am
>
> 1) <u>BST</u> is in force so we add on one hour   .... 11.30 am
>
> 2) <u>Longitude</u> is 1.25 deg. west of Greenwich
>   so we add 1.25 X 4 = 5 minutes   .... 11.35 am
>
> 3) From the <u>sundial time conversion graph</u>
>   we add 6 minutes for this date.
>   So this gives   **clock time** ...... 11.41 am

Sundials
further activity 1

Now it is your turn.

Use an atlas or an ordnance survey map to find the longitude of the place where your sundial is.
Try to read this to the nearest quarter of a degree.

Work out the clock time for the following times and dates using the longitude for the position of your sundial.

DATE	SUNDIAL TIME	CLOCK TIME
May 25th	2.30 pm	
February 14th	9.00 am	
September 1st	12.00 noon	
November 17th	3.00 pm	

It may be that you want to know what your sundial **should** read at a certain time by your clock. This is particularly useful when you are setting up a dial and you do not know for certain the north-south direction. Indeed, this is quite a good way of finding the north-south direction. All that you have to do is to use the same method that you used before, but backwards. One example is given to show you how it is done:

Example: What time should a sundial be showing in longitude 1.25 degrees west on 15th August when a clock shows 4.30 pm?

clock time : 4.30 pm

From the sundial time conversion graph (p.33) *subtract* 5 min.   -0.05
For longitude west *subtract* 1.25x4 min.   -0.05
For BST *subtract* 1 hour   -1.00

sundial time : 3.20 pm

Sundials
further activity 2

Dial hunting

You will need :-

> **A detailed map of the area**
> **A protractor**
> **A ruler, pencil and notepad**
> **A packed lunch**

Whether you live in the town or the country, a dial hunt is an enjoyable way to spend a day out of school in the holidays or at weekends.

Start with a visit to your local library, they often have papers written by members of a local history society which may mention a sundial. The town hall or a Heritage Centre should be able to help with locating old buildings or with finding dials in public places.

The oldest buildings around are often churches so these are good places to start.
A sundial could be on any wall but a south facing wall will get the most sunshine so this is the best place to look. Churches are usually aligned east - west with the tower ,or spire, at the west end, so you should be able to find the south wall, but check with the sun.
Look on the church tower, dials were often placed there in earlier times. Most were replaced with clocks in the last century,but some remain and some new dials have been made in recent years, there are four on the tower of St. Margaret's, Westminster.
Look in the churchyard, sundials were often carved on gravestones or used as memorials and, inside the church, a sundial can form part of a stained glass window.

Public gardens or parks sometimes have sundials, originally to tell the time but also as memorials, or simply because they are both beautiful and interesting.
Large old houses, such as National Trust properties, often have sundials either on the wall of a building or as a garden ornament.

Not all dials are old, modern buildings sometimes have new ones and many beautiful dials have been made in this century. The National Maritime Museum at Greenwich, London, has an interesting collection of old and new dials and a modern sundial park is being added to the Visitors Centre at Jodrell Bank, Cheshire.

When you find a sundial there are a great many things to record about it. Copy the form on the next page and take a supply with you.
Completed forms can be sent to Mrs. J. Walker (address on the back cover) and any 'new' dials you find will be added to the national register.
If you are able to take a photograph or make a brass rubbing of the dial plate please include a copy.

Sundials
further activity 2

Name _____ Age _____ Date _____

School (include the address) _____

Type of dial (Compare with the photographs on page 17 and tick the correct box)

horizontal ☐   vertical ☐   equatorial ☐   polar ☐

armillary sphere ☐   analemmatic ☐   other ☐

Situation   eg.   Bodnant Garden, Tal-y-Cafn, Colwyn Bay, Clwyd.
In the garden at the west side of the house.

[ ]

Grid reference if possible [ ]

Dimensions   If the dial is accessible measure in mm. If it cannot be measured give an estimate.

dial plate   diameter [ ] mm   circular dials only

length [ ] mm   width [ ] mm   rectangular dials

height above ground [ ] mm

gnomon   angle it makes with the dial plate [ ] degrees

length of gnomon [ ] mm

Time marks   circle the hours which are marked on the dial.

am - 3 4 5 6 7 8 9 10 11 12   pm - 1 2 3 4 5 6 7 8 9

Are the numbers Roman ☐   or Arabic ☐

Other marks   eg. compass points or signs of the zodiac

[ ]

Materials used   wood, stone, brass, bronze, etc.

[ ]

Inscription   Copy any writing on the dial (continue on the back of the form).

[ ]

Dial maker   Copy the maker's name and the date if given.

[ ]

Sundials

group project 1

Bicycle Wheel and Broomstick Sundial

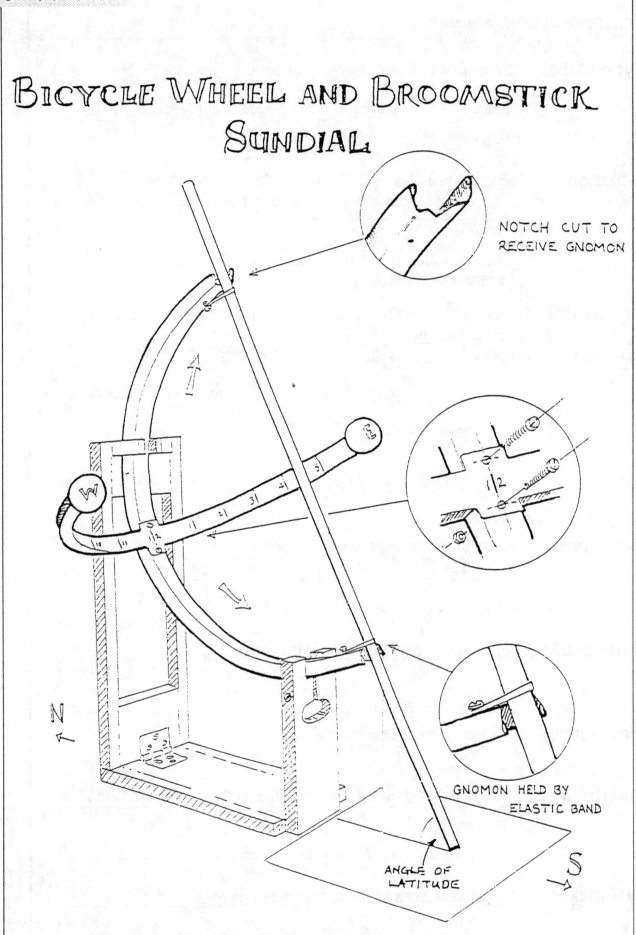

Sundials

group project 1

Bicycle wheel and broomstick dial

This equatorial type sundial was designed for use in schools with four main criteria in mind:

> *Inexpensive*
> *Easy to construct*
> *Quickly assembled / dismantled*
> *Requires little storage space*

The Dial

The main frame of the dial consists of a large cycle wheel rim cut exactly in half. The two halves are then reassembled laid across each other at right angles to the centre of both. One half forms the meridian (north-south) band and the other the equatorial (east-west) band.

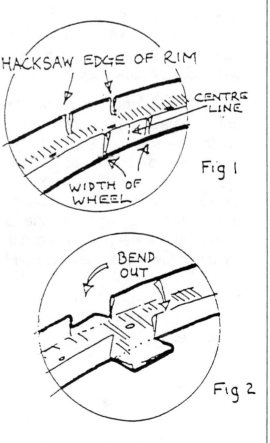

To secure the two bands together, hacksaw the rims of the equatorial band in two places equidistant from its centre just wide enough to equal the width of the wheel (Fig.1). The two small tabs between these sawcuts can then be bent outwards at right angles to their original positions (Fig.2). The two halves of the rim should fit over each other quite neatly. The two small tabs may require a little dressing with a hammer to ensure that they are quite flat.

Sundials

group project 1

Drill a small hole in the centre of each tab and corresponding holes in the other half of the rim and insert two small screws with nuts to fix both halves together (Fig.3).

At both ends of the meridian band cut out a V-notch to the centre of the rim (Fig.4). This will accommodate the gnomon which is made from a length of wooden dowel or broom handle. The gnomon is secured to the main frame by elastic bands which are looped over a screw placed in the nearest spoke hole (Fig.5).

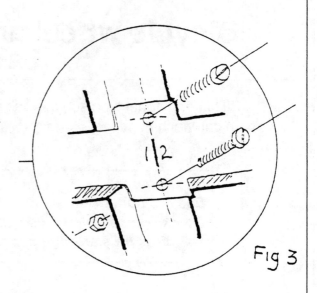

Fig 3

The ends of the equatorial band should be covered to prevent cuts from sharp edges. Table-tennis balls, cut open, can be pushed over the ends to provide a satisfactory finish. These can be decorated with the appropriate letters E and W.

The equatorial band is marked out in hour lines every 15 degrees from the centre line and numbered with the hours of the day. 12 noon is at the centre of the dial with the afternoon hours extending along the E quadrant and the morning hours along the W one. Smaller marks can be inserted between the hour lines to show the half and quarter hours.

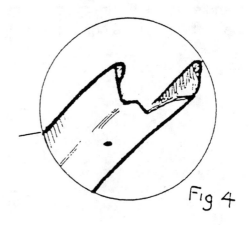

Fig 4

The main body of the dial is now complete.

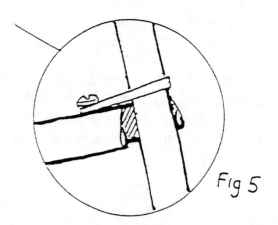

Fig 5

Sundials

group project 1

The cradle

The wooden cradle illustrated has been designed to enable the dial to be adjusted to the angle of latitude by rotating the meridian band through the cradle. The dial can easily be set to this angle by placing a card or plywood template, cut to the angle of latitude, under the bottom end of the gnomon as in the main illustration.

Once the dial is set to the correct angle the meridian band is secured by a clamp screw in the shorter leg. This template not only demonstrates to the pupils the importance of setting the gnomon to the angle of latitude but can also be used as a traditional horizontal sundial. The hour lines can be marked out using information from the table on page 65.

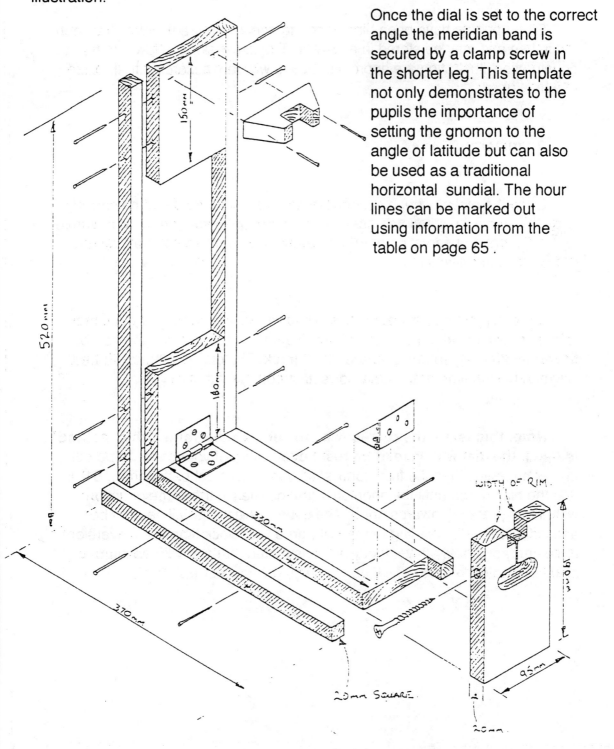

Sundials

group project 1

Constructing the cradle

The cradle is made from standard planed timber available from DIY shops and timber merchants. Most of the construction is simple nails-and-glue carpentry.

Both uprights have been hinged so that the whole cradle folds flat when not in use and may be stored in a drawer. The dimensions shown in the construction illustration are approximate and will vary according to the size of wheel used.

Alignment

Once the dial has been assembled and set in the cradle at the correct angle it can be oriented in an open position on horizontal ground. For simple time-telling and observation of the sun's apparent movement, the dial may be set up using a watch.

For example, if your watch says 10.30 when setting up your dial, then rotate the whole structure until the centre of the shadow of the gnomon on the equatorial ring comes over the 10.30 mark. The dial, if left undisturbed, will then tell the time reasonably correctly throughout that day.

Note: this is not the correct way to set up a dial, and if this method is used, the dial will have to be reset every day. If your dial is to be set up permanently, then the true north-south direction must be found (p. 61) and the dial turned until the gnomon of the dial lies in the same direction with the higher end towards north. The usual corrections will need to be applied to each dial reading to make it agree with clock time. A conversion table or graph could conveniently be constructed, but a quick and simple solution is to make a sundial time conversion calculator (p.30).

Sundials

A ' bicycle wheel' equatorial dial with its projection onto a horizontal surface (dark) and an analemmatic dial (light).

Sundials

group project 2

An analemmatic dial

This is ideal for playgrounds. There are no parts which can get broken or lost, or which can be at all dangerous.

An analemmatic dial is shown on the front cover of this book.

The dial is elliptical and uses hour points rather than hour lines. A child can stand at the appropriate date with one foot on each side of the central line and read the time from his or her own shadow.

The dial may be painted on a playground, or set out on a lawn using small paving slabs or bricks for hour markers and larger ones for the central date scale. A 'trial run' chalking the dial on to tarmac will help get the technique right.

For first-school-sized children, the distance AB should be about 5 metres, so the semi-major axis M will be 2.5 metres (Fig.1).

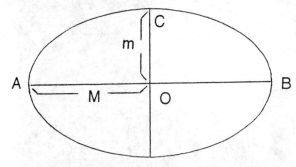

M = semi major axis

m = semi minor axis

Fig 1

<u>To calculate m (the semi-minor axis)</u>

Find the angle of latitude of the site to the nearest quarter of a degree using an Ordnance Survey map.

Find the sine of this angle from a scientific calculator or a book of tables.

Then m = M x the sine of your latitude.

e.g. for a dial 5 metre across at latitude 52 degrees,

   m = 2.5 x sine 52

   = 2.5 x 0.7880

   <u>m = 1.97 metres</u>

Sundials

group project 2

Drawing the ellipse

1. Find the 'clock time' which gives a 'sundial time' of noon (see p.32).

2. Using a plumb line suspended above the point which is to be the centre O of your dial, or a vertical stick in soft ground at O, mark in the line of its shadow at ' sundial noon'. (See also 'Finding south' p. 61).

3. Draw a line of length equal to m from O towards north. Call the end of this line C. Draw a line at right-angles to OC through O, and mark the points A and B which are both 2.5 metres from O. All of these points can be marked with masonry nails hammered into tarmac, or with pegs or meat skewers in soft ground.

4. Make a loop of non-stretchy string equal (in doubled length) to M. Loop the string round C and use it like a pair of compasses to mark off the points X and Y on the major axis AB. Drive in two more pegs at X and Y (Fig. 2).

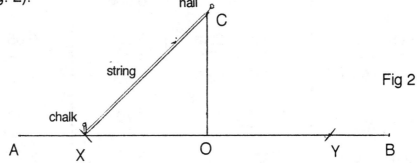

Fig 2

5. Make another loop of string so that it fits tightly round the three pegs at X, Y and C. Remove the peg at C.

6. Place a piece of chalk or suitable scratching marker inside the loop and, keeping the string taut, draw the ellipse through A, B and C. Continue for a little way beyond A and B (Fig. 3).

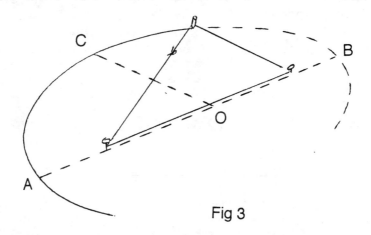

Fig 3

Sundials

group project 2

Marking the hours

Table 1 gives hour values (H) for hours of the day from 6am to 6pm GMT or from 7am to 7pm BST. (A dial can be made with both of these included but it would be best to choose one or the other to start with.) These hour values H are the sine of the angles 0, 15, 30 degrees etc.

H x M is the hour distance for each hour of the day. The hour distances are given for a dial with a major-axis of 5 metre.

Table 1 : hour distances for an analemmatic dial

Time		H	hour distance H x M metre
GMT	BST		
12noon	1pm	0.000	0.00
11am 1pm	12noon 2pm	0.259	0.65
10am 2pm	11am 3pm	0.500	1.25
9am 3pm	10am 4pm	0.707	1.77
8am 4pm	9am 5pm	0.866	2.17
7am 5pm	8am 6pm	0.966	2.41
6am 6pm	7am 7pm	1.000	2.50

Measure the hour distances out from O, on both sides, along AOB (Fig.4). Mark the hour points on the ellipse by drawing lines at right angles to AOB from each of the hour distances.

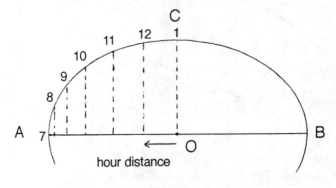

Fig 4

Sundials

group project 2

Marking out the central scale of dates

All the distances on this scale are marked out from the centre point O of the dial. The distances D are calculated from D = M x Z where M is the length of the semi-major axis and Z is given by Z = (tan dec) x (cosine latitude).

Table 2 gives the angle of declination (dec.) of the sun for the first day of each month (this is the angle the sun is above or below the celestial equator).
Also given is the tangent of the declination (tan dec).
The value of Z is given for latitude 52 degrees, but can be found for any latitude using the formula above.

Table 2 : Values for the central scale (lat 52 deg)

Date	dec (deg.)	tan dec	Z	D (cm)	Date	dec (deg.)	tan dec	Z	D (cm)
Jan 1	-23.0	-0.424	-0.261	-65.3	Jul 1	+23.1	+0.427	+0.263	+65.7
Feb 1	-17.1	-0.308	-0.189	-47.4	Aug 1	+18.0	+0.325	+0.200	+50.0
Mar 1	-7.5	-0.132	-0.081	-20.3	Sep 1	+8.2	+0.144	+0.089	+22.2
Apr 1	+4.6	+0.080	+0.050	+12.4	Oct 1	-3.2	-0.056	-0.034	-8.6
May 1	+15.1	+0.270	+0.166	+41.5	Nov 1	-14.5	-0.259	-0.159	-39.8
Jun 1	+22.1	+0.406	+0.250	+62.5	Dec 1	-21.8	-0.400	-0.246	-61.6
*Jun 21	+23.4	+0.433	+0.266	+66.6	*Dec 21	-23.4	-0.433	-0.266	-66.6

*approximate date of solstice

The distance to be measured along OC for each date is D. Positive values are measured towards C, negative values away from C (Fig.5). If a child stands with feet astride the central line OC at the correct date, the time can be read from the place where their shadow cuts the ellipse of hour points.

Note: Although it is possible to build in a correction for longitude, this has not been done. To all the times shown by the dial there should be added 4 minutes for every degree of longitude west, or subtract 4 minutes for every degree of longitude east.

Sundials

group project 2

Notes

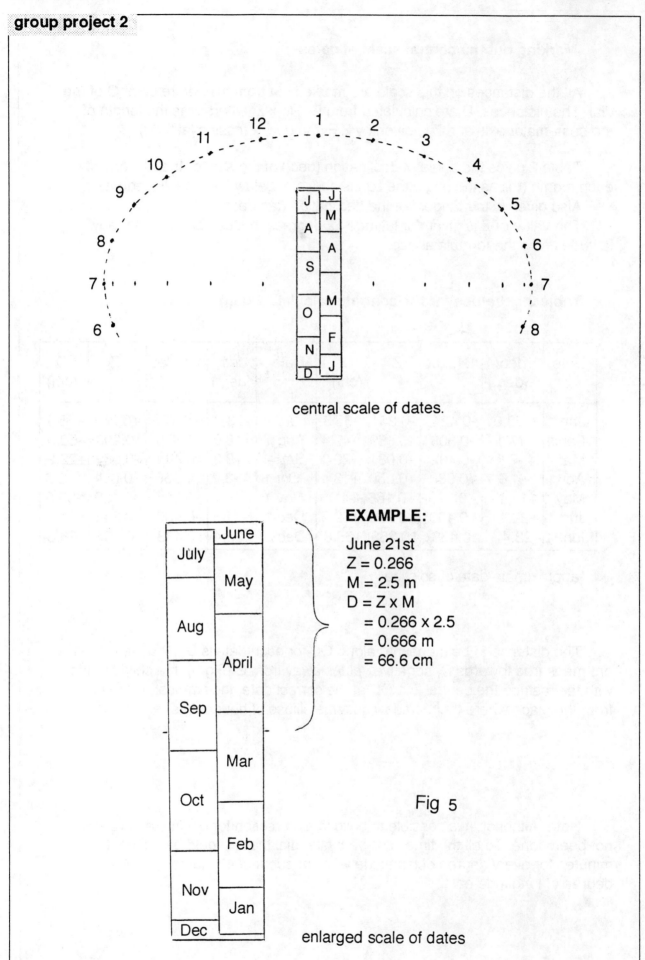

central scale of dates.

EXAMPLE:

June 21st
$Z = 0.266$
$M = 2.5$ m
$D = Z \times M$
  $= 0.266 \times 2.5$
  $= 0.666$ m
  $= 66.6$ cm

Fig 5

enlarged scale of dates

Sundials

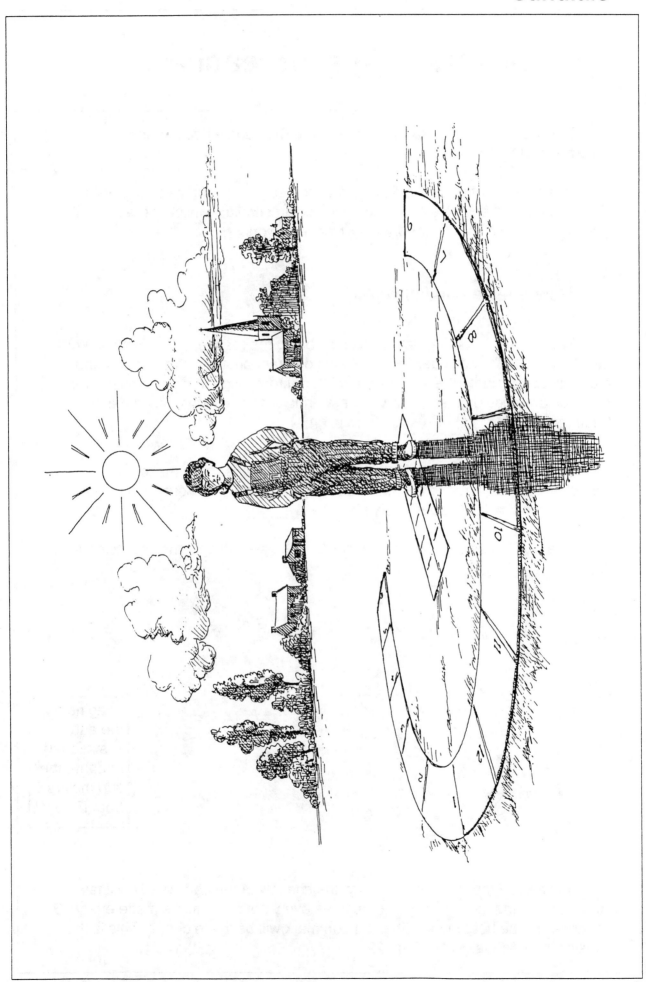

Sundials *Notes*

General notes for teachers

This section of the book is intended to provide some background information for teachers.

It is hoped that by providing a variety of designs for sundials the pupils will find interest and enjoyment in making and using sundials whilst at the same time gaining some insight into the way in which they work.

How does a sundial work?

The answer is too frequently given as being the response to "How do you use the sundial?" Each question requires a different answer - the first demands much greater insight - the second is a matter of following the instructions! The notes below should go some way to answering both questions. Much deeper treatment is given in the texts listed on p. 68.

All sundials make use of the fact that the sun *appears* to make one complete revolution round the earth (360 degrees) every 24 hours. So it seems to move through 15 degrees every hour.

Imagine the earth to be hollow with its axis as a rod right through its centre.

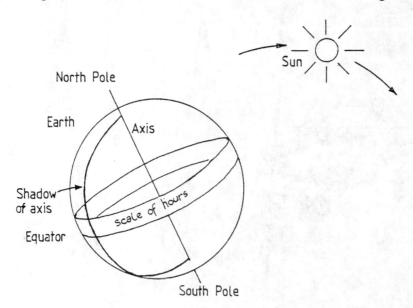

Imagine that the earth is stationary and that the sun moves from E to W

As the sun moves across the sky, the shadow of the earth's axis will fall on the equator and move through 15 degrees every hour. So marks made every 15 degrees will be hour marks. The 12 noon mark will be in the centre. This is the basis for the equatorial dial (p. 22).

Notes   Sundials

Instead of having a band round the equator to catch the shadow of the axis (which is really called the 'gnomon') we could just as well have a disk in the plane of the equator.

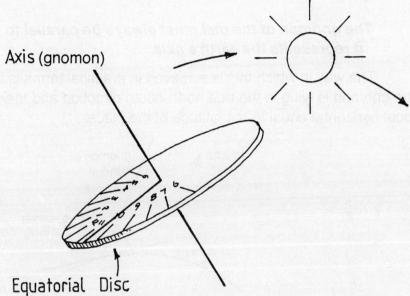

The hour marks are established in exactly the same way as before, and can be drawn on the upper (north) face of the equatorial disk as well as the lower (south) face.*

This gives rise to another form of equatorial dial and is used for the model on p. 20 .

---

*The reason why there needs to be a scale on both upper and lower faces of this equatorial disc is that in summer months (approx. March 21 - Sept 21) the sun appears to move across the sky above the plane of the equator, and below it during the winter months. This is caused by the tilt of the earth's axis to the plane of the earth's orbit round the sun (the 'ecliptic').

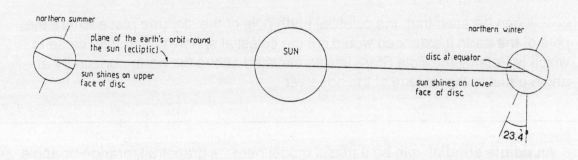

Sundials — Notes

So our sundials are no more or less than miniature versions of the earth and its axis! But this brings out a very important feature of sundials:

The gnomon of the dial must always be parallel to the axis of the earth - it represents the earth's axis.

The way in which this is achieved in practical terms is to make sure that the gnomon is lying in the true north-south direction and makes an angle to the local horizontal equal to the latitude of the place.

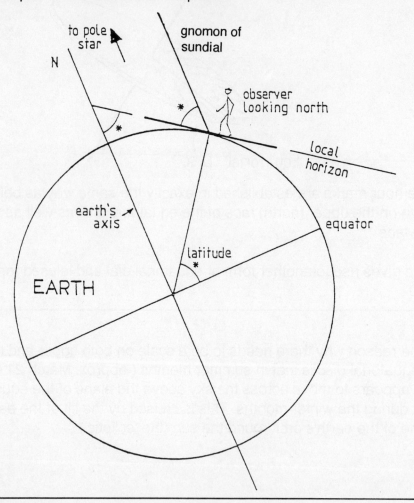

It can be seen that the celestial north pole of the sky (the place where the pole of the earth if extended would cut the celestial sphere) - and very close to which is Polaris, the Pole Star - is also elevated above the north horizon by an angle equal to the latitude of the observer.

An **edible sundial** can be a useful model here - a grapefruit, orange or apple can act as the earth, and a barbecue stick or knitting needle pushed through its centre acts as the earth's axis. A chocolate orange would do equally well, with the added advantage of equal angles around the equator being provided by the chocolate segments!

Notes   Sundials

Of course, the sundial does not have to be an equatorial band or an equatorial disc. Many common dials are horizontal or vertical. This simply means that the 'plate' of the dial (the surface on which the hour marks are made) is horizontal or vertical.

The positions of the hour lines on these dials can be derived from an equatorial one by projection.

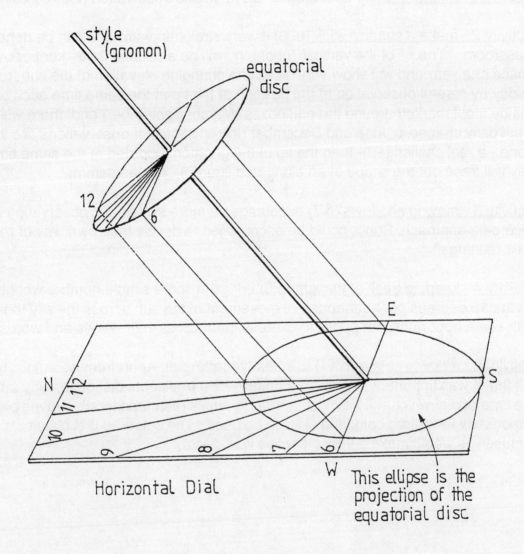

Vertical dials are derived in the same way, but this time the equatorial disc is projected on to a vertical surface.

Vertical dials may face exactly due south and are called 'vertical direct south' dials. If they face in any direction other than south, they become 'vertical declining' dials, and the angle their face is turned away from South to the East or West is called the 'declining angle'. Some of the books listed on p. 68 go into the details of how these dials are designed - but *all of them have their gnomons aligned north-south and parallel to the earth's axis.*

Sundials   Notes

Notes on activities 1 to 5

<u>Be back in time</u> (p. 4) is aimed at getting pupils to think about the movement of shadows caused by the sun being associated with the passing of time.

<u>Activity 1- shadows</u> (p. 5) encourages a systematic observation and recording.

<u>Activity 2 - make a shadow stick</u> (p. 6) is very straightforward and can be done in the classroom. The *tip* of the vertical 'gnomon' will be a reliable time - keeper over the space of a year and will show very nicely the changing elevation of the sun from day to day by careful observation of the position of the tip at the same time each day. This will be most marked around the equinoxes (March, September) and there will be least daily change in June and December (the solstices). If observations are very well done - a real challenge(!)- then the tip of the gnomon recorded at the same time each day will trace out the shape of an elongated figure 8 - the 'analemma'.

<u>Activity 3 - moving shadows</u> (p.7) encourages a more systematic observation and recording approach. Pupils could be encouraged to devise their own way of recording their readings.

<u>Activity 4 - keeping track of the sun</u> (p.8) brings in some simple number work based on the 15 degrees per hour apparent movement of the sun across the sky together with some opportunity for pupils to express themselves in drawings and words.

<u>Activity 5 - a dial to colour</u> (p.11) is a 'ready-made' dial. Apart from its simplicity it was felt that it was important to provide a pattern for a traditional dial quite early on. This is the type of dial which will be most familiar to pupils (and teachers!). It gives plenty of opportunity for artistic colouring skills to be used. The coloured dial plates themselves would make a very attractive wall display.

Notes *Sundials*

Notes on projects 1 to 6   (see p13 for photographs of finished dials)

Project 1 - a horizontal dial (p14) is what 'everyone expects' a sundial to look like and is therefore a good starting point. If strong card or plywood is used a lasting and reliable sundial can be made.

Project 2 - a horizontal dial from a matchbox (p 18) .These are very quick and easy to make and are surprisingly accurate. They have been modelled on the standard matchbox sizes such as England's Glory.
Project 2 can be completed in a few minutes, (before the sun goes in!) and does not need any measurement of angles.
Dividing the sides of the box by eye was thought to be easier than measuring the distances with a ruler, and produces equally satisfactory results for this type of dial. If the match is positioned as in Fig.3 it will make an angle of very nearly 52 deg. with the dial plate which is satisfactory for most of England. Obviously, some accuracy in time-keeping has been sacrificed to simplicity of manufacture. The greatest errors are at 9am and 3pm when the dial will be slow by 17 minutes (the width of the matchstick), but the other hour lines are accurate to within 5 minutes.
If dividing by eye is a problem for younger children,the actual size diagram of Fig.2 on p19 could be photo-copied and stuck on to the back of the matchbox tray.

Project 3 - an equatorial dial (p20)
The distances have been calculated so that the straw will be inclined at
52 deg. to the horizontal. Ideally the angle should equal the latitude of the place where the dial is to be used. This angle can easily be checked with a protractor, or a cardboard template could be made for the use of the whole class. For other latitudes use the table on p65.
The dial should be positioned so that the straw is pointing due north.
If the dial has been correctly made and correctly positioned then the semicircular dial plates will lie in the plane of the earth's equator and therefore of the celestial equator. The sun crosses the celestial equator at the equinoxes (approx. 21 March, 21 Sept.). On these days the sun rises due E and sets due W (ignoring small corrections which are not necessary for this accuracy of work). The shadow of the straw should fall equally on both scales of the dial. In the summer months, only the scale on the upper plate will be illuminated because the sun is above the plane of the celestial equator (its *'declination'* is north (positive)). In the winter months, only the scale on the lower plate will be illuminated because the sun is below the plane of the celestial equator (its declination is south (negative)).

Notes   Sundials

Project 4 - an equatorial dial from a washing up liquid bottle (p.22)
The 'seams' from the manufacturing process give the vertical edges conveniently at opposite ends of a diameter.

Marking the hour lines on plastic can be a problem. The plastic must be dry. A fine-pointed permanent ink overhead-projector pen gives a good result.

When the dial is complete the gnomon (knitting-needle, etc) should be at an angle to the horizontal equal to the user's latitude. Hanging from the finger, as shown in Fig.3 p.23, will give this angle fairly accurately, but greater precision can be gained with a little experimentation. For instance, if the upper horizontal cut is made somewhat lower than the 2 cm suggested, say 4 cm, then by cutting a V-notch at the centre of this upper edge, and testing the resulting hanging-angle, then trying again, the correct angle can eventually be arrived at.

For further ideas on this project see p. 57

Project 5 - a polar dial from a matchbox (p. 24)
The hour lines could just as well be drawn by eye on the inside of the matchbox tray, but this involves several stages of division. The paper-folding method should lead to more accurate results.

Project 6 - a polar dial made from wood (p.26)
The two pieces of wood can be of any reasonable size as long as the length of the gnomon is equal to the width of the dial plate and the length of the dial plate is at least nine times the height of the gnomon.

Notes   Sundials

Extension work on project 4

There are several refinements which can be made to this simple idea:

1. The bottle tends to move about if there is any wind. The effect can be overcome by using a permanent mounting such as by embedding the bottom of the bottle in plasticine or clay.

2. If a coat hanger is used as the gnomon, the protruding ends can be fashioned to form a support for the dial (Fig. 6, p.58). Alternatively, the wire could be pushed through a wooden base (Fig. 7 p.58). **Make sure that there are no sharp ends of wire protruding from anywhere!**

3. A more elaborate support, which removes the need for the needle or wire is to construct a gnomon of card which is self-supporting (Figs 8, 9). Cut a slot in the back of the bottle along the 12 noon line and insert the tongue which must be equal in depth to the radius of the bottle. The free edge of the tongue should be parallel to and along the axis of the bottle. The bottle will need some extra support at its base using plasticine, for example, to keep it centrally positioned on the tongue.

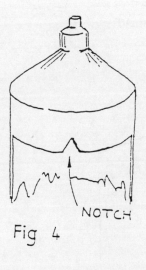

Fig 4

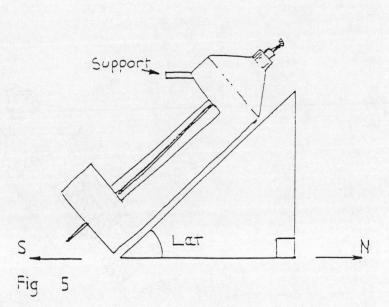

Fig 5

Sundials

Notes

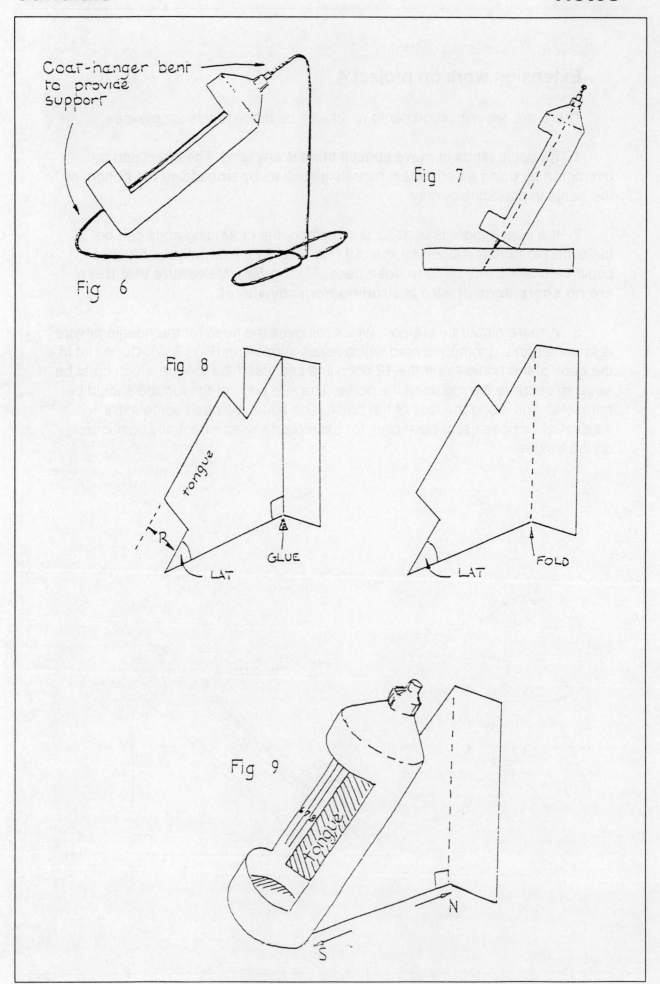

Notes Sundials

Further ideas on project 4

The washing-up liquid bottle could equally-well be replaced by a large pop-bottle. To remove the label of the bottle, keep the *outside* dry, but pour hot water into the bottle. This will melt the adhesive sufficiently to allow the label to be peeled off.

The graduations inside the bottle need not be as shown in fig.3 p.23 provided that the longitudinal cuts are straight and at exactly opposite ends of a diameter. The two straight edges thus formed can then act as gnomons, the one on the eastern side for the morning hours, and that on the western side for the afternoon hours. Only six graduations are needed - a small amount of elementary geometry will soon show why.

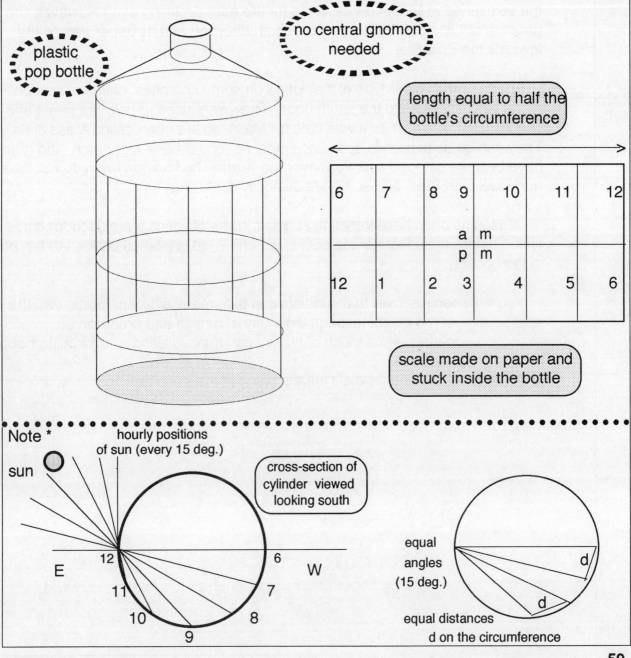

Sundials   Notes

Dial Hunting (p.36)

An obvious starting point is the church. Before the advent of Greenwich Mean Time clocks kept local time. Very often the parish church was the place where a large sundial was put up so that mechanical clocks and watches could be regulated against the sun. It almost goes without saying that before you take a class to a church, you should make sure that there is a sundial there for the pupils to see. Lists are available from B.S.S. (see page 2).

Look for dials high up on the south side of the tower or south-facing wall. You may find that the dial is of the vertical declining type because the wall is not facing exactly south. (With a west-declining dial, the gnomon is to the east of the vertical noon line, and vice versa for the east-declining dial.) However, the gnomon still lies exactly in the north-south direction and its upper end points towards the Pole Star.

Another much older type of dial found on some churches, called a 'scratch dial', is usually close to the south door. They were used to help the priest and congregation know when it was time for Mass, so are often called Mass dials. They can be quite hard to spot since most no longer have a gnomon, and many have become defaced and worn over the years. The lines on them do not follow our usual time lines, and some are distinctly confusing!

Museums often have very fine sundials some of which were used for time-telling before the advent of the pocket watch - so are small enough to be put in the pocket.

You will soon discover that sundials can take many different forms, and the appreciation of the combination of ingenuity of design and precision of execution, together with a touch of history can make dialling a fascinating hobby.

In addition, hunting for dials means being in the open air!

Finding South

All sundials have to be aligned to the compass points NSEW.

If the direction of true south is known the rest can easily be found.

It is not usually a good idea to rely on 'local knowledge' as to where true south is because a few degrees out can lead to large errors in time shown by the dial.

There are several ways in which south can be found:

(a) Use a compass.
   This is not a reliable method, nor accurate, but it is quick.
   Whitaker's Almanac gives full details of how to allow for changes in the direction of the earth's magnetic field.

(b) Cheat a little and use the sundial itself by setting it (on a sunny day!) to what it should read at that moment.
   Remember to make the usual corrections to change sundial time to clock time.

(c) Use the method below which does not require any corrections.

(d) Follow other methods referred to in books on dialling
   (see booklist on p.68).

Equipment

> **A point at a fixed height above a horizontal surface**
> **A plumb line**
> **A spirit level**
> **A ball of string**
> **Some pegs such as meat skewers**

The fixed point can be provided by a rigid pole such as a bamboo cane or broom handle, or even the roof corner of a building - anything whose shadow can be traced throughout the day on horizontal ground (or nearly so) (Figs. 1,2). The point could be the tip of a long nail or rod driven at right angles through a piece of white plastic-coated board from a DIY shop (Fig.3).

Sundials

Notes

finding south

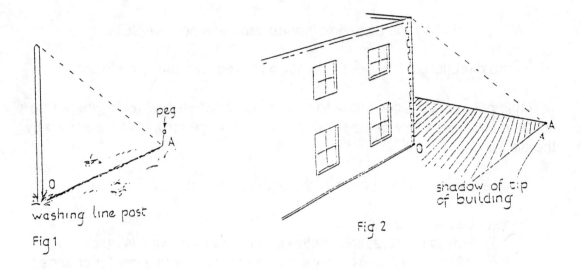

Fig 1 — washing line post, peg, A, O

Fig 2 — shadow of tip of building

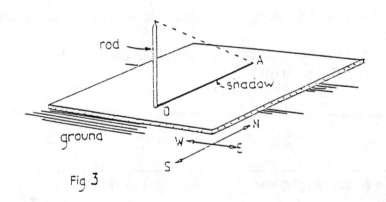

Fig 3 — rod, shadow, ground, A, O, N, W, E, S

Mark the positions of A, the tip of the shadow, for several hours before and after noon GMT (1pm BST). Chalk can be used on hard ground, and wooden pegs or meat skewers on soft ground. Pencil or felt-tipped pen can be used on plastic coated surfaces. It is not necessary to put marks at hourly intervals, or even on the hour, but the more often A is marked throughout the day, the better. The periods of about three hours before and after local noon are the most important.

Notes
Sundials

finding south

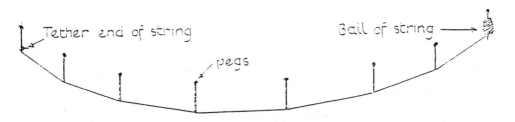

Shape obtained depends on the time of year

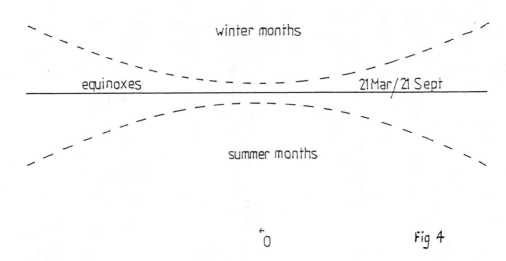

Fig 4

When the marking of the positions of A has been completed, the points need joining with a smooth curve. The more points you have marked, the easier this will be. Where pegs have been placed in the ground a piece of string can be stretched gently along them and held in place as a gardener does. The shape of the curve depends on the time of year (Fig.4).

Next find the point O vertically below the tip of the pole etc. which has been giving you points A. You may need to use a plumb line for this. With O as centre and using a pair of compasses (or a length of non-stretchy string) draw a circular arc which cuts the path (locus) of A in two places as far apart as possible. Call the intersections P and Q (Fig.5).

Bisect the angle POQ by drawing arcs centred at P and then Q (the same radius as each other) to cross at R (Fig.6).

Join R,O and this is the north-south line. Mark it in some permanent way. It may also be useful to join P,Q which is the east-west line. As a check, you should also find that the north-south line cuts the locus of A where the shadow was shortest, and this occurred at the moment of **local noon.**

Sundials

finding south

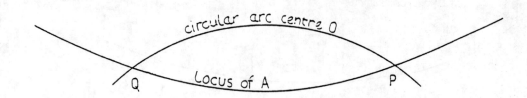

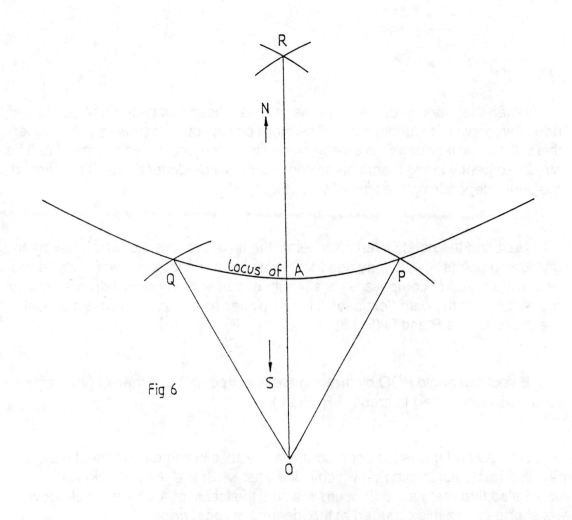

Fig 5

Fig 6

Notes *Sundials*

Tables for simple dials

Table 1: Angles for hour lines on horizontal dials

Hour / Latitude	12	11 & 1	10 & 2	9 & 3	8 & 4	7 & 5	6
49	0	11	24	37	53	70	90
50	0	12	24	37	53	71	90
51	0	12	24	38	53	71	90
52	0	12	24	38	54	71	90
53	0	12	25	39	54	71	90
54	0	12	25	39	54	72	90
55	0	12	25	39	55	72	90
56	0	13	26	40	55	72	90
57	0	13	26	40	55	72	90
58	0	13	26	40	56	72	90
59	0	13	26	41	56	73	90

Table 2: Conversion of longitude to time (in min.)

Longitude	2E	1E	0	1W	2W	3W	4W	5W	6W	7W	8W
Correction (apply to dial reading)	-8	-4	0	+4	+8	+12	+16	+20	+24	+28	+32

Sundials

Glossary

Analemmatic dial. A type of horizontal dial which uses a vertical, pin shaped, gnomon to cast the shadow. The gnomon must first be set to the correct date.

Armillary sphere. A sundial made up of circular hoops, one of which indicates the time by the shadow of the central axis. The dial on page 38 is an incomplete form of armillary sphere.

Axis. The line about which a body revolves. The earth's axis is a line from the north pole, through the centre, to the south pole.

B.S.T. Stands for British Summer Time. In many countries clocks are put forward one hour during the summer months.

Datum line. A line from which other lines and angles are measured. For sundials the datum is generally the 12 o'clock line.

Declination. The angle of the sun, north or south of the equator, as it would appear from the centre of the earth. This varies from zero degrees at the equinoxes to a maximum of 23.44 degrees (north or south) at the solstices. Its value on any particular day can be found from tables.

Dial plate. The surface of a sundial on which the hour lines are marked.

Ecliptic. If you imagine the sun and the earth to be floating in a bath of water, the surface of the water is equivalent to the plane of the ecliptic. The planets are also going round the sun in planes not far removed from this surface.

Equatorial dial. A sundial whose dial plate is designed to be parallel to the plane of the earth's equator. See the diagram on page 51.

GMT. Stands for Greenwich Mean Time. A 24 hour time system based on the Greenwich meridian and previously used as a world-wide time reference. Now replaced by Universal Time (UT), but most people still use the old expression.

Gnomon. The part of a sundial which casts the time shadow on the dial plate. Strictly speaking, the edge of the gnomon which gives the shadow is called the **style** but, to avoid confusion, the word gnomon is used throughout this book.

Horizontal dial. A sundial whose dial plate is horizontal. See page 13, photograph 1, for an example.

Sundials

Latitude. Lines of latitude can be found drawn on a globe as circles, parallel to the equator, becoming smaller and smaller as they get nearer to the poles. The equator is at latitude *0 degrees* and the poles are at latitude *90 degrees* these being the angles they make when measured from the centre of the earth. You can also find these imaginary lines drawn on an atlas. Latitude is used, among other things, for setting the angles of gnomons (see page 21). **Lat** is the abbreviation for latitude.

Longitude. The lines on a globe which pass through both poles and cross the equator at right- angles. They are marked along the equator in degrees east or west of the Greenwich meridian (which is 0 degrees longitude), the angles again being measured from the centre of the earth. **Long** is the abbreviation for longitude.

Mean Time. A time based, not on the real sun, but on an imaginary sun supposed to cross the sky at a constant speed at all seasons of the year. This gives hours of exactly equal length, summer and winter, unlike the hours found from a sundial which vary slightly from day to day.

Meridian. An imaginary arc in the sky starting from above the north pole, crossing the equator at right-angles, and finishing above the south pole.
A **local meridian** is the arc which goes overhead at a particular place, eg. the Greenwich meridian passes over the old observatory at Greenwich and follows the line of 0 degrees longitude.

Noon. When the sun appears to cross a local meridian it is noon, or midday, at that place. (Note that this is sundial noon , not necessarily 12 o'clock by your watch.)

Polar dial. A sundial whose dial plate is parallel to the polar axis (see page26).

Solar day. The interval of time between the sun crossing a meridian one day and crossing it again the next day. About 24 hours but sometimes a few seconds more or less depending on the time of year.

Sundial. An instrument for telling the time by the sun.

Variation. Magnetic compasses rarely point exactly to true north. The angle between the compass north and true north is known as the variation and varies from place to place. Sundials are always set to true north.

Vertical dial. A sundial with its dial plate upright. The only vertical dials considered in this book are those which face due south but they can be designed to face in any direction ------- even due north !

Booklist

Books for younger readers:

Anno Mitsumasa **The Earth is a Sundial** Bodley Head 1985 £10
0 370 31016 0   A pop-up book for stimulating early interest.

Jenkins **Sundials and Timedials** Tarquin Books 1987 15pp and 8 models £4.00 0 906212 59 6 Basic theory and cardboard cutouts of working dials and devices using the sun.

Books for older readers:

Cousins Frank W. **Sundials, A Simplified Approach by Means of the Equatorial Dial** John Baker 1969.
Now out of print, but well worth trying to get from your library.

Daniel C.St.J. **Sundials** Shire Pubs. (No.176) 1986 32pp £1.75
0 85263 8086
Historical introduction, with illustrations of many different types of dial. Very little theory. No constructional details.

Mayall & Mayall R.N.& M.W. **Sundials** Sky Publishing Corp., Cambridge, Mass. 1938 repr.1973, 1989 250pp £10.00 0 933346 60 3
Assumes no previous knowledge - covers much the same ground as Waugh but has a very different approach. Construction details given.

Mills Robert **Things for Sky Watchers to Do** Pheon Books 200pp 1991 £9.75 1 873347 00 6
Enthusiastic elementary astronomical text with a useful section on the Sun.

Vincent Carole **Time and the Sundial** Carole Vincent 1988 30pp £1.00
Theory and background to the Plymouth Armada Way / New George Street dial.

Waugh A.E. **Sundials, Their Theory and Construction** Dover Pubs. 1973 228pp £4.50 0486 22947 5.
Comprehensive and authoritative. Drawing and calculation methods.

Notes   Sundials

books

Specialist booksellers who may be able to obtain these books are those such as:

    Rogers Turner Books   Mrs R.K.Shenton,
    22 Nelson Road,   148 Percy Road,
    Greenwich,   Twickenham,
    London SE10 9JB   London. TW2 6JB
    081 853 5271   081 894 6888

Molehill Press of Grange Farmhouse, Geddington, Kettering, Northants NN14 1AL ,issue science packs for schools, including "Instrument Pack for Earth in Space (AT16)", which has designs for sundials etc. and background information.

Percy Seymour; 'Adventures with Astronomy' Murray 1983
ISBN 07195 3931 5 has some information on sundials.

Christopher St. J.H. Daniels ' Stained Glass Window Sundials'
Country Life Feb26 1987 shows another rare form of sundial.

The Association for Astronomy Education with Association for Science Education have produced 'Earth and Space' - a workpack for Primary and Middle Schools 1990 0 86357 132 8 covering all astronomical topics in the national curriculum levels 1-5.
ASE: College Lane, Hatfield, Herts AL 10 9AA,
AAE: c/o The London Planetarium, Marylebone Rd., London NW1 5LR

Thanks

The Education Group of The British Sundial Society wishes to record its thanks to those schools which have helped in reviewing the initial draft of this book and who have used some of the materials. We are most grateful for their comments and suggestions and have tried to incorporate the ideas in the revision.

Kingswood School, Bath
France Hill School, Camberley, Surrey.
Priors Court, Chieveley.
Sherard School, Melton Mowbray
St. Julian's Primary, Wellow, Bath.
Ravenscroft Primary, Jaywick, Clacton-on-Sea.
Finmere School, Finmere, Bucks.

Sundials Notes

British Sundial Society Addresses

Dial Registrar	Mr Gordon E. Taylor Five Firs Cinderford Lane Cowbeech Hailsham E. Sussex BN27 4HL	0323 833255
Membership	Mrs Janet Thorne 15 Chesterfield Rd Laira Plymouth Devon PL3 6BD	0752 227582
Education	Mrs Jane Walker 31 Longdown Rd Little Sandhurst Camberley Surrey GU17 8QG	0344 772569
Secretary	Mr David A. Young Brook Cottage 112 Whitehall Rd Chingford London E4 6DW	081 529 4880

Places to visit - where sundials can be seen, and much more besides.

Jodrell Bank Visitors Centre, Macclesfield, Cheshire   0477 71339

The Maritime Museum, Greenwich, London   081 858 4422

The Science Museum, South Kensington, London   071 938 8000